Simon Schwörer

Aus der Reihe: e-fellows.net stipendiaten-wissen

e-fellows.net (Hrsg.)

Band 571

The Human Genome Project

GRIN Verlag

Bibliografische Information der Deutschen Nationalbibliothek:

Die Deutsche Bibliothek verzeichnet diese Publikation in der Deutschen National-
bibliografie; detaillierte bibliografische Daten sind im Internet über http://dnb.d-
nb.de/ abrufbar.

Imprint:

Copyright © 2012 GRIN Verlag GmbH
Druck und Bindung: Books on Demand GmbH, Norderstedt Germany
ISBN: 978-3-656-31468-4

This book at GRIN:

http://www.grin.com/en/e-book/204136/the-human-genome-project

GRIN - Your knowledge has value

Der GRIN Verlag publiziert seit 1998 wissenschaftliche Arbeiten von Studenten, Hochschullehrern und anderen Akademikern als eBook und gedrucktes Buch. Die Verlagswebsite www.grin.com ist die ideale Plattform zur Veröffentlichung von Hausarbeiten, Abschlussarbeiten, wissenschaftlichen Aufsätzen, Dissertationen und Fachbüchern.

Visit us on the internet:

http://www.grin.com/

http://www.facebook.com/grincom

http://www.twitter.com/grin_com

The Human Genome Project

Essay

seminar Bioethics

Simon Schwörer

08/21/2012

▶1◀

"For the genetically superior, success is easier to attain but is by no mean guaranteed. After all, there is no gene for fate. And when for one reason or another, a member of the elit falls on hard times, their genetic identity becomes a valued commodity for the unscrupulous." [1]
  — Vincent Freeman, character in the movie Gattaca.

In the movie Gattaca, a future is depicted where parents may design the genetic equipment of their offspring prenatally. Since not everyone has access to this technology, individuals without genetical enhancement are considered "in-valids". Although it is illegal to discriminate on the basis of genes, profiling a person's genotype is common and used to qualify for employment. While only the "valids" are allowed to climb the social ladder, the invalids, considered more susceptible to disease and a shorter lifespan, are relegated [1, 2].

Gattaca warns of the problems that arise if humans are reduced to nothing more than their genes, an ideology called genetic determinism. In the movie, the society generally accepts genetic determinism, culminating in the view that a genetic read-out acts as unchangeable prophecy, as predestination [3].
How likely is such a future, where one's chances in life are restricted by his or her genetic code?

Among other scientists, Francis Collins, the director of the National Human Genome Research Institute (NHGRI) at the National Institutes of Health (NIH) in the United States, attended a screening of the film [3]. Knowing the *"science fiction buzz"* about human genetics was important to him and should also be for other researchers in the Human Genome Project (HGP) [3].
And he knew what he was talking about: already years before the movie, the HGP has been source of similar ethical concerns as in Gattaca.

▶2◀

"The problem [with genetic research] is, we're just starting down this path, feeling our way in the dark. We have a small lantern in the form of a gene, but the lantern doesn't penetrate more than a couple of hundred feet. We don't know whether we're going to encounter chasms, rock walls or mountain ranges along the way. We don't even know how long the path is." [4]
  — Francis S. Collins, director of the American NIH and the HGP (1990).

Starting in 1986 as a Human Genome Initiative by Charles DeLisi of the United States Department of Energy (DOE), the project received its final boost in 1988 by the enthusiastic effort of James Watson, predecessor of Francis Collins as head of the NHGRI [5]. Eventually, the HGP was founded in 1990 as DOE/NIH joint-project with funding of about $200 million a year and was expected to take 15 years [5].

Today, the achievements of the HGP can be seen as outstanding; it created an infrastructure for biology and medicine [6]. But then, the idea of sequencing the entire 3 billion bases of the human genome was discounted as "absurd", "dangerous" and "impossible" by numerous critics [5]. And they were not that wrong: in 1990, the technology for whole genome sequencing was just not available. Therefore, sequencing the genome was postponed in the HGP until better technology has been developed that would make it faster and cheaper. To ensure that these technologies were soon available, some large-scale sequencing projects were funded by the HGP [6].

The first goal of the HGP was to construct high resolution linkage maps of polymorphic markers of the human chromosomes [6]. By linkage analysis, the arrangement of genes on the chromosomes can be determined upon co-inheritance of polymorphic DNA markers, but it provides now way to actually locate the genes on the chromosomes. Thus, physical maps of overlapping cDNA fragments covering the whole genome were build concurrently to eventually locate the markers from the linkage map on the DNA. Thereby, the identification of disease genes was made easier [6].

Besides, another focus of the HGP was to analyze and ultimately sequence the genomes of simple model organisms like the bacteria Escherichia coli (E. coli), the fly Drosophila melanogaster or the mouse Mus musculus [6]. This was done for several reasons: first, even if one day a possible disease gene has been encoded, the function of the resulting protein cannot be deduced just from its amino-acid sequence only [6]. But since evolution has been quite conserved between different organisms, a human gene might have its homologue gene in a model organism which allows figuring out that gene's function much more easily. Second, learning to sequence the best is easier on a smaller genome of a model organism than on the large human genome [5].

But as the first real "big-science" project, the HGP also terrified the scientific community to a certain extent: people feared that on the one hand, money was pushed away from smaller projects and on the other hand, the creativity and freedom of individual researchers were dramatically reduced by this targeted and highly organized effort. For the HGP founders, it was just that the need for more coordination and organization in the project was exceeding what people were accustomed to [6]. To drop these arguments, the HGP early tried to maximize the public benefit by making available new tools to the scientific community and the release of new data before publication [7].

Apart from the sequence data and genomic as well as physical maps of the human genome, the project also produced some rather surprising findings. First, the genome sequence between different people is almost identical - 99.9% of the nucleotide bases are the same. Second, less than 2% of the genome encodes for proteins. And third, the total number of genes, previously estimated at 80,000 to 140,000, is only around 23,000. Unfortunately, the function of more than the half of them is yet unknown [8].

In addition, the HGP's key goals also included addressing the ethical, legal, and social issues (ELSI) that may arise from the project. Therefore since its inception, 3% to 5% of the annual HGP budget was devoted to study the implications of the rapidly increasing knowledge about the human genome and the technological progress in this field on individuals and society. Thereby, the HGP ELSI program represents the world's largest bioethical program. Among other issues, ELSI topics include: fairness in the use of genetic information, privacy and confidentiality, commercialization of products as well as psychological impact and stigmatization [9].

In the following, I will present the most challenging ELSI topics. Some of them are not directly related to the HGP, but definitively emerge from the increasing the availability of genetic information to which the project immensely contributed.

►3◄

"We used to think our fate was in our stars. Now we know, in large measure, our fate is in our genes." [10]

— James Watson, Nobel prize winner for the discovery of the DNA structure (1989).

"Each individual is entitled to lead a life in which genetic characteristics will not be the basis of unjust discrimination or unfair or inhuman treatment." [11]

— Human Genetics Commission (2002).

In Gattaca, a genetic read-out after birth determines the destiny of the main character Vincent: with predispositions for mental disorders and heart failure as well as a reduced life expectancy, he is a victim of genetic discrimination right from the start [2]. Although *"it's illegal to discriminate on the basis of genetics -genoism it's called- [...]"* [1]. Vincent seizes the only chance to avoid genetic discrimination by "borrowing" the genetic profile of an injured star athlete [2]. This is just a movie, but how real is such a genetic discrimination nowadays?

Basis of this concern is always a genetic test, by which a person's DNA - taken from cells in a sample of blood or, occasionally, from other body fluids or tissues - is examined for some anomaly that represents a disease or disorder. This anomaly may be large, like a missing or additional peace of a chromosome, or small like a missing, additional or alternated nucleotide base. For instance, genetic testing is used to confirm suspected DNA mutations responsible for a certain phenotype or to look for a possible predisposition for a disease. Today, around 1,000 genetic tests are available [12].

I think many people are still believing that genetic discrimination is more science-fiction than reality, but in the United States, cases have already been reported where employers used findings from genetic tests to avoid recruiting candidates that were shown to possess genes associated with susceptibilities for diseases that are expensive, interminable or incurable [13]. Besides employers, people are afraid that also insurance companies might take use of genetic information to calculate the risk of developing certain diseases for their candidates. As result, individuals with high risk status might be totally excluded, have reduced coverage or pay astronomically high contributions for a health or life insurance.

To counter this development, a law called GINA became effective in 2008 in the United States [13]. This "Genetic Information Nondiscrimination Act" prohibits American insurers and employers *"from discriminating on the basis of information derived from genetic tests"* and additionally, they are not allowed to demand a genetic test [13]. This is a first step, but still it remains to be generally declared who should have access to personal genetic information and who not, to prevent the misuse of that sensitive information. In my opinion, this knowledge is highly confidential.

So if genetic information may influence an employer or an insurer for the choice in their candidates, how could it affect an individual itself? Suppose a genetic test results in a very high probability for a yet incurable disorder like Huntington's disease, what would be your reaction? How would you behave? How would you continue living you life? No one can answer these questions without having been in this situation himself. And the fact that genetic testing most often only provides a probability and not a certainty makes life not easier. Some people who carry a disease-associated mutation never develop the disease.

Besides, another issue would be how to keep that information private; how to avoid the people around you from noticing it? Genetic information is highly confidential, and once it becomes public, it implicates a high risk of stigmatization, even if the changes are not visible at first glance. Thereby the society's perception of the affected individual would change, even if most people would certainly argue against it.

Not only after birth or as adult, genetic testing is available also prenatal. Pre-implantation genetic diagnosis (PGD), also known as embryo screening, allows for instance prenatal sex discernment and may therefore be used to select embryos in one sex preferentially. One may use it to screen for a specific genetic disease; it is available for many monogenic disorders such as cystic fibrosis or Huntington's disease. Especially for pregnant women of advanced age, PGD is used as a screening for chromosomal abnormalities in the embryo and thereby may increase the rate of pregnancy success, as in this case pregnancy loss due to aneuploidy is a frequent problem [14, 15]. Recently, it has been published that even whole genome sequencing of the embryo from blood samples of pregnant women is possible [16].

"Designing babies" may be an option of PGD as well, including for example the modification of physical characteristics of the unborn child by changing its genetic composition [17]. This may be one step towards eugenics as in Gattaca and raises several questions: is modifying the genetic code for non-therapeutic reasons ethically justifiable? Should "imperfect" embryos or ones with gene defects be aborted? Or would such a practice even be legally allowed in the future?

In Germany, PGD is generally prohibited, but since the end of 2011, it is allowed if a severe genetic disorder in the child or miscarriage is possible due to the genetic predisposition of the parents [18] .

Currently there are no regulations for evaluating the accuracy, reliability and utility of genetic tests [9], so one cannot definitively know if they are safe and the results trustable. Some genetic tests were developed in pubic laboratories, while others are marked by companies as kits that can be purchased by everybody [9]. This is critical since certainly not everybody is able to interpret the results or will engage genetic counseling. A lot of education for the handling of genetic information is hence necessary, especially for healthcare personal [9].

But in this whole debate, one may not forget that genetic testing has saved many lives. It may identify individuals at risk for preventable conditions or be of help for a diagnosis and kick-start the appropriate treatment.

What is more, research on the human genome made great progress in the last years, and we not only know a lot about single gene disorders, but also complex conditions like cancer, Multiple sclerosis or Alzheimer's disease could be successfully linked to multiple genes and gene-environment interactions. However, there still not any prospect for curing these diseases in the near future. The current lack of available medical options is still subduing the technological advance. Therefore, should genetic testing actually be performed if no treatment is available?

One could argument that with the knowledge "you will die in ten years", some people would enjoin their remaining lifetime, for example do crazy things which they always wished to do. In contrast, other people might rather be burdened themselves with the oncoming disease and would therefore retrospectively prefer not knowing about their condition. This argument established the Gene Diagnostics Law (GenDG) in Germany in 2009: genetic tests can only be performed with the consent of the respective individual [19].

▸4◂

"This is part of what makes us who we are. ... How can they claim information that is in all of our bodies?" [20]

—— Jesse Reynolds, policy analyst for the Center for Genetics and Society (2010).

Another issue that arises from the availability of the human genome is patenting and commercialization of the same. Generally, a patent confers on its proprietor a temporally limited monopoly for an invention, an individual property right. To be patentable, the subject must be novel, inventive and industrially applicable. In genetics, this involves the isolation of the respective gene, the specification of the resulting product and the determination of its function in nature, thereby providing the industrial applicability [21].

Nowadays, there are already about 1,000 genes that have been patented in the United States, and millions of genome-related applications have been filed [22].

Although the HGP early announced to publish the human DNA sequence in the freely available database GenBank of the National Center for Biotechnology Information (NCBI), the concern of patenting DNA sequences was long time feared, as there was a head-on-head competition in the late stages between the public HGP and the private venture Celera to sequence the whole human genome [5]. Celera was under Craig Venter's command, who worked for the NIH in the first days of the HGP. The company planned to release the data only on its web page and not in a public database, and additional analysis may not be gratitude [5]. Despite a lot of dispute, both would eventually achieve simultaneous publications in Nature and Science, respectively [5].

Still, there is the question "who owns genes"? Genes were not invented by the companies who sequence or map them; they are like a common heritage of mankind, not a property of a scientist or a company [22]. Of course, patenting a gene would be the researcher's reward for his discovery and he could surely use the money he gains from the patenting for his research [22]. But licensing the rights to use gene sequences is quite disturbing. Certainly, their accessibility would be limited and thus, the time for development of diagnostics and therapeutics would be dramatically enhanced, as further research on the human genome is definitely required for therapeutic purposes [22].

Last but not least I want to make a short excursion to some philosophical implications of the "new genetics" that were introduced by the HGP. Inevitable in this context is the debate of free will versus genetic determinism [9]. *"Free will is a philosophical term of art for a particular sort of capacity of rational agents to choose a course of action from among various alternatives"* [23]. Contrarily, genetic determinism is *"the theory that human character and behavior are shaped by the genes that comprise the individual's genotype rather than by culture; environvent; and individual choice"* [24]. Believing in this concept raises the question how someone can be held responsible for something if we are but victims of our genes. But do genes make people really behave in a particular way [9]?

Imagine a gene encoding for traits like intelligence or criminality would be discovered and genetic diagnostic of these traits would be available. This would certainly shatter our society; our whole social and legal system would collapse overnight.

But no single gene determines a particular behavior. Behaviors are complex traits involving multiple genes that are affected by a variety of other, environmental factors [25].

Within this consideration, what kind of influence does a gene really have? The only true is that they are responsible for our general physical traits.

A rather controversial thesis on the purpose of genes and their "behavior" is the book *The Selfish Gene* by Richard Dawkins, published in 1976. This book on developmental biology refers to humans as *"survival machines, robot vehicles blindly programmed to preserve the selfish molecules known as genes" [26]* (p. 2). Dawkins attributes the evolution of life to the selection of those genes able to generate the most copies of themselves: *"They are in you and in me; they created us, body and mind; and their preservation is the ultimate rationale for our existence" [26]* (p. 24). With this he argues for genetic determinism; people only behave like they are doing to maximize the number of copies of its genes that are passing on globally [26]. After publication, the book rapidly became popular. Even before many basic genetic mechanisms were understood, Dawkins could complete the Darwinian theory of evolution to a certain extent [27]. Although is provoked many controversies, *The Selfish Gene* is still an important contribution to sociobiology, which studies the social behavior of all creatures.

>5<

"We are here to celebrate the completion of the first survey of the entire human genome. Without a doubt, this is the most important, most wondrous map ever produced by human kind." [28]
    — Bill Clinton, ex-U.S. president (2000).

"Mapping the human genome has been compared with putting a man on the moon, but I believe it is more than that. This is the outstanding achievement not only of our lifetime, but in terms of human history. [...] This code is the essence of mankind, and as long as humans exist, this code is going to be important and will be used." [29]
    — Michael Dexter, director of the Wellcome Trust (2000).

In 2000, the first draft sequences of the genome were released, and finally the complete human genome was published in 2003, two years ahead of schedule [5].
The sequences released by the HGP had a quick impact on finding genes associated with disease. But draining meaningful knowledge about the provided DNA sequence still is a big project for the next years, if not decades. But new technologies enable researchers to approach questions about the interconnection between genes, transcripts and proteins systematically and in large scale, leading to faster and more detailed findings concerning human genetics [8].

Being now able to completely sequence the human genome gave rise to new, basically similar projects. One of these aims to fully sequence the genomes of several human cancer cells, a project called The Cancer Genome Atlas (TCGA) [30]. It was launched in 2006 with funding from the

NHGRI and the National Cancer Institute (NCI) and posses an enormous and complex effort, as Francis Collins described it, *"the equivalent of more than 10,000 Human Genome Projects [...]"* [30]. Designed as limited pilot project, the TCGA rapidly became a large scale operation and already yielded first important data, like the characterization of new mutations common in many cancers and the identification of 11 to 15 crucial pathways typically disrupted in cancer [30]. As with the data provided by the HGP, this information will hopefully convert into better understanding and treatment of disease(s).

Although the HGP was completed, especially the progress in the field of full genome sequencing did not stop. Year in and year out, technology improves and makes this task easier, cheaper and faster. The trend seems unstoppable. Craig Venter's genome was one of the firsts that were nearly completely sequenced, and he even published a book telling the story of the race for the sequence of his genome [31]. Apple founder Steve Jobs had his genome sequenced for $100,000 four years ago [32]. For $299, the company "23andMe" offers testing for thousands of Single Nucleotide Polymorphisms (SNPs), thereby providing genetic risk estimation for certain diseases and discovery of ancestry composition [33].

In 2011, a paper describing the first successful use of whole exome sequencing for the diagnosis and treatment of a patient was published [34], and moreover the company Illumina, one of the market leaders, announced to charge only $4,000 for sequencing the whole human genome [32], and whole exom sequencing is with around $700 even cheaper. The still ongoing price decline will enable a more rapid growth in the field of personalized medicine and the discovery of rare genetic disorders. With decreasing costs, access to the related technologies will not stay limited only to the rich. According to Illumina's CEO, Jay Flatley, *"by 2019 it will have become routine to map infants' genes when they are born"* [32]. Perhaps we are already on our way towards a future as depicted in Gattaca.

▸References◂

1.   *Gattaca*, 1997. Niccol, A., DVD, 106 min, USA: Columbia Pictures.
2.   Wikipedia, the free encyclopedia, *Gattaca*, 2012. http://en.wikipedia.org/wiki/Gattaca (2012/07/15).
3.   Kirby, D.A., *The New Eugenics in Cinema: Genetic Determinism and Gene Therapy in GATTACA*, 2000. http://www.depauw.edu/sfs/essays/gattaca.htm (2012/07/15).
4.   Nash, M.J., *Tracking Down Killer Genes.* TIME Magazine, 1990/09/17.
5.   Roberts, L., *The human genome. Controversial from the start.* Science, 2001. **291**(5507): p. 1182-8.
6.   *Mapping the Genome: The Vision, the Science, the Implementation.* Los Alamos Science. **20**: p. 68-93.
7.   Collins, F.S., Morgan, M., and Patrinos, A., *The Human Genome Project: lessons from large-scale biology.* Science, 2003. **300**(5617): p. 286-90.
8.   U.S. Department of Energy Office of Science, *Insights Learned from the Human DNA Sequence*, 2009.
     http://www.ornl.gov/sci/techresources/Human_Genome/project/journals/insights.shtml (2012/07/15).
9.   U.S. Department of Energy Office of Science, *Ethical, Legal, and Social Issues*, 2011.
     http://www.ornl.gov/sci/techresources/Human_Genome/elsi/elsi.shtml (2012/07/15).
10. Jaroff, L., *The Gene Hunt.* TIME Magazine, 1989/03/20.
11. Human Genetics Commission, *Inside Information: Recommandations*, 2002.
     http://www.google.de/url?sa=t&rct=j&q=each%20individual%20is%20entitled%20to%20le
     ad%20a%20life%20in%20which%20genetic%20characteristics%20will%20not%20be%20t
     he%20basis%20of%20unjust%20discrimination%20or%20unfair%20or%20inhuman%20tre
     atment.&source=web&cd=6&ved=0CGEQFjAF&url=http%3A%2F%2Fwww.hgc.gov.uk%
     2FUploadDocs%2FContents%2FDocuments%2Fiirecommendations.pdf&ei=QL8CUNH_L
     4XptQaLuP3ZBg&usg=AFQjCNGyZfjZQDRdIFi_zT_IIgXzWglxtA&cad=rja
     (2012/07/15).
12. U.S. Department of Health and Human Services, *Understanding Gene Testing*,
     http://www.accessexcellence.org/AE/AEPC/NIH/index.php (2012/07/15).
13. U.S. Department of Energy Office of Science, *Genetics Privacy and Legislation*, 2008.
     http://www.ornl.gov/sci/techresources/Human_Genome/elsi/legislat.shtml
14. Reproductive Genetics Institute, *What is PGD*, 2011.
     http://www.reproductivegenetics.com/pgd.html (2012/07/15).
15. U.S. Department of Energy Office of Science, *Gene Testing*, 2010.
     http://www.ornl.gov/sci/techresources/Human_Genome/medicine/genetest.shtml (2012/07/15).
16. Kitzman, J.O., et al., *Noninvasive whole-genome sequencing of a human fetus.* Sci Transl Med, 2012. **4**(137): p. 137ra76.
17. Future Human Evolution, *Genetic Engineering*, 2010.
     http://www.humansfuture.org/genetic_engineering_designer_babies.php.htm (2012/07/15).
18. ZEIT online, *Bundestag erlaubt Embryonenauswahl im Labor*, 2011.
     http://www.zeit.de/wissen/2011-07/pid-bundestag-entscheidung (2012/07/15).
19. BfDI, *Gendiagnostikgesetz*, 2010.
     http://www.bfdi.bund.de/DE/Themen/GesundheitUndSoziales/Gesundheit/Artikel/Gendiagn
     ostikgesetz.html (2012/08/20).
20. Shelton, D.L., *Who owns your genes?* Chicago Tribune, 2010/04/01.

21. Gleiter, H. and Riepe, H.G., *Seminar "Introduction to Patent Law" for Molecular Medicine*. 2012.
22. U.S. Department of Energy Office of Science, *Genetics and Patenting*, 2010. http://www.ornl.gov/sci/techresources/Human_Genome/elsi/patents.shtml (2012/07/15).
23. O'Connor, T., *Free Will*, 2002. Stanford Encyclopedia of Philosophy (2012/07/15).
24. *Genetic Determinism*, encyclopedia of medical concepts (2012/07/15).
25. U.S. Department of Energy Office of Science, *Behavioral Genetics*, 2008. http://www.ornl.gov/sci/techresources/Human_Genome/elsi/behavior.shtml (2012/07/15).
26. Dawkins, R., *The selfish gene*. 1976, New York: Oxford University Press. xi, 224 p.
27. Wikipedia, the free encyclopedia, *The Selfish Gene*, 2012. http://en.wikipedia.org/wiki/The_Selfish_Gene (2012/07/15).
28. *Remarks on the completion of the first survey of the entire Human Genome Project*, 2000/06/26. TV broadcast, USA: The White House, Office of the Press Secretary.
29. *The first draft of the Book of Humankind has been read*, 2000/06/26. TV broadcast, USA: Wellcome Trust and the Sanger Centre.
30. Mukherjee, S., *The emperor of all maladies : a biography of cancer*. 1st Scribner hardcover ed. 2010, New York: Scribner. xiv, 571 p.
31. Venter, C., *A Life Decoded - My Genome: My Life*. 2008: Penguin Books. 400p.
32. Wikipedia, the free encyclopedia, *Whole genome sequencing*, 2012. http://en.wikipedia.org/wiki/Whole_genome_sequencing#cite_note-14 (2012/07/15).
33. 23andMe, Inc., *Your DNA, Endless Possibilities*, 2012. https://www.23andme.com/ (2012/08/20).
34. Worthey, E.A., et al., *Making a definitive diagnosis: successful clinical application of whole exome sequencing in a child with intractable inflammatory bowel disease*. Genet Med, 2011. **13**(3): p. 255-62.